ACCESS TO MATH

RATIOS AND PROPORTIONS

TEACHER'S MANUAL

GLOBE FEARON EDUCATIONAL PUBLISHER
A Division of Simon & Schuster
Upper Saddle River, New Jersey

Printed in the United States of America 2 3 4 5 6 7 8 9 10 99 98

ISBN 0-8359-1498-4

GLOBE FEARON EDUCATIONAL PUBLISHER
A Division of Simon & Schuster
Upper Saddle River, New Jersey

Contents

MEETING STUDENT NEEDS

Most classrooms include students with a wide variety of different needs and learning styles. As you plan your lessons, you can include teaching strategies that address the needs you have identified. Although the ideas that follow are stated in terms of specific needs, all students can benefit from these strategies.

ESL/LEP STRATEGIES

Mathematical learning is made up of three elements: the mathematical concept, its symbolic notation, and the vocabulary with which these elements are expressed. The ESL student will benefit from a constant focus on the interrelationship among these elements. Whenever you present a mathematical concept, stress the connection between the language and the concept by giving simply worded definitions and specific examples. When you present a symbolic notation, stress the connection between the language and the symbols both orally and in written form. Since this approach can be helpful to native speakers also, it will be advantageous to the entire class.

A variety of techniques can help strengthen students' math and language skills.

- Assign ESL students to a small group within which they can feel comfortable and receive individual attention. If there are several students whose first language is the same, assign them in pairs to a group. Add a native speaker of English who wants to help others and who will be patient and understanding. Alternatively, partner ESL students with native speakers at all stages of learning, from the conceptual stage to practice exercises.
- Use a combination of written and spoken representations of terms and mathematical symbols. Always connect spoken terms with what they look like in writing, and vice versa.
- Regularly write essential vocabulary on the board and say these words aloud. Also ask students to say and write the words.
- When necessary, rephrase definitions of vocabulary or math terms in several ways. If students cannot grasp the definition phrased in one form, it may reach them more efficiently when worded another way.
- Encourage students to paraphrase definitions in their own words. This increases language facility and also lets students restate the ideas in terms with which they are comfortable.
- Continually give specific examples of concepts or terms. Whenever possible, use examples from everyday life. Draw on students' previous knowledge and experience. Also ask students to give their own examples.
- Whenever necessary, model concepts and processes with concrete materials or with diagrams. Also encourage students to use manipulatives and diagrams.
- Have students create and use flash cards that feature vocabulary words, math symbols, and other useful mathematical elements. Set aside time for flash card work on a regular basis.

COOPERATIVE LEARNING

Success with cooperative learning groups depends on teaching students, in incremental steps, how to work together. Employers consider the ability to solve problems as part of a team a key skill in the people they hire. Additionally, many at-risk students learn best in a cooperative learning setting. These two facts alone make it worthwhile to spend time teaching students how to work effectively and stay on task in a group situation.

Begin by familiarizing students with the actual process of forming groups quickly and efficiently. Start with partners or groups of three, of mixed abilities. Teach students how to move desks or chairs quickly and quietly. Insure individual participation by assigning specific roles within the interdependent group. For example, roles might include summarizer, checker, reader, researcher, recorder, and so forth. As students become more familiar with the process, they can often choose, assign, and title their own roles.

Assign a mathematical task to be accomplished by the group. Explain that the group will have accomplished the task when each member can answer four out of five problems correctly. Allow plenty of time and explain that finishing first among group members (a competitive rather than a cooperative approach) is not the point. Rather, the point is for the group to work together to ensure that all members master the skill. You may want to test mastery at a second class meeting, after a short review. If a group does not meet the criteria, ask members to work together to clear up misconceptions and confusions. Encourage those who have succeeded to coach and motivate others who have not yet been successful.

Some students may have difficulty buying into group work and group accomplishments. Tell them that teamwork is a realistic reflection of the way work is done in the world outside the classroom. Help quicker students to realize that the best understanding of a concept comes from teaching it to someone else. Assure them that they are internalizing their own knowledge into long-term memory when they explain a concept to another student.

The following tips will help you create a beneficial environment for learning math cooperatively.

- Allow the task to determine group size. Groups of 2 or 3 might work to understand or review a process such as dividing fractions. Groups of up to 6 might formulate problems or work on solving complex word problems.
- Keep group meetings and tasks short at first and gradually lengthen them.
- Before work begins, give students a clear statement of what they need to know and do to complete the assignment. This information should be written on the chalkboard or on handouts. Include the following:
  a. mathematical objectives (what the group will learn, what skills and processes the group will use, and what product or solution the group should expect to arrive at)
  b. collaborative objectives (how the group will function to achieve its goals)
  c. the amount of time that will be given to complete the assignment
  d. the criteria by which the work will be evaluated (for example, will the group have to explain the process it used? Come up with a single, accurate answer? Give a range of possible answers?)
- Assign necessary roles, such as recorder or checker, at the start of the activity.
- Monitor the groups' progress. When necessary, intervene to answer questions, reinforce math skills, or stress appropriate group procedures.
- Talk about and teach conflict management skills to help students resolve differences. Remind them that they will often have to work with people they did not choose, but that they can still get the job done.
- Assess results in terms of both groups and individuals. When possible, have groups compare their results with those of other groups.

Manipulatives

The term *manipulatives* is used to describe any objects students can handle and work with as they learn mathematics. Multiplication grids, chips, counters, and coins are all useful manipulatives for many mathematical concepts.

All students learn better and remember longer when they have concrete experiences that relate to abstract ideas. Research tells us that at-risk students have a higher proportion of tactile and kinesthetic learning styles than the student population in general. Thus, these students learn best if they are introduced to a topic in ways that involve touching and moving objects (tactile) and moving their large muscles (kinesthetic).

Manipulatives can be used to demonstrate concepts and processes in many areas of intermediate and advanced classrooms, as they are in primary grades. Of course, adjustments should be made for the different situation. For instance, the age level of the students should be considered when choosing materials. Also, because upper-level classrooms might not have budget allocations for "official" manipulatives such as place value blocks, everyday materials such as paper clips or even liquids can be used. Here are some tips for working with manipulatives:

- For many situations, uniform size and shape is essential. Paper clips may be useful. Paper is easily obtained and can be cut into any size and shape. Dried beans and pasta are inexpensive even in large quantities. (However, the use of food items as manipulatives may be frowned on in certain communities, such as some Hispanic communities, as wasteful.)
- In some situations, color coding is helpful. Marbles of different colors make good choices.
- In situations where parts of a whole must be modeled and perhaps reshaped or removed, consider modeling clay, kneaded erasers, dough made of flour, salt, and water.
- Measuring implements such as marked measuring cups may be useful for investigating fractions or decimals.
- A balance can be useful for more than just measurement activities. The concept of making sure that both sides of an equation are equal in value can be demonstrated with pennies and a balance. An unknown quantity can be shown with pennies wrapped in paper. (Caution: New pennies and old pennies have slightly different weights, so pre-weigh pennies to be sure they are uniform.)
- Play money can be useful for both money computation and place value work. Bills and coins can be made from paper and cardboard if premade play money is not available.
- Diagrams can be useful as a substitute for manipulative objects in some situations.
- Students themselves can become "manipulatives," acting out certain processes and substitutions.

LESSON PLANS

LESSON 1

Introducing the Lesson

Give students the following problem: Twelve students walk into a cafeteria where there are four tables. They sit down at the tables so that the same number of students is at each table. How many students are there per table? What does the word "per" mean in the previous question?

Teaching the Lesson

A ratio is a comparison of two quantities. Often, being able to work with such a comparison gives us a context for a quantity that is more meaningful than just seeing the quantity by itself. For example, the simple fact that there are 185 member nations in the United Nations doesn't give us much of an indication of what part of the world is represented in the organization. But knowing that there are 185 member nations to 71 nonmember nations makes it clear that there are many more member nations than nonmember nations, and that even though the UN roster includes most nations of the world, it's not even close to including *all* nations, and so on. As students develop their sense of the size of numbers and their sense of just what a ratio is, they can begin to refine the knowledge that they can take from the ratio $\frac{185}{71}$: there are more than twice as many member nations as nonmember nations, more than $\frac{2}{3}$ of the world's nations are represented in the UN, and so on. These facts cannot be gleaned from the simple statistic that there are 185 UN member nations.

A rate is a special type of ratio, one that compares two quantities with different measures. Explain to students that while a comparison of UN members to nonmembers is not a rate because both terms in the ratio represent numbers of countries, a comparison of countries to years is a rate. Rates are often used in the sense of matching up objects from two groups—for example, matching up 3 countries per year, 1 house for every 2 families, or 6 occurrences in every hour. You may want to point this out to students when reviewing with them the different words that are used to indicate ratios. Also point out that the words are not all interchangeable and that they often give clues as to the nature of the relationship being described. For example, while "to" may be used to express comparisons of different subsets within a larger set (such as 185 members to 71 nonmembers), "out of" is used to compare a subset to a set or a part to a whole (such as 185 members out of 256 nations).

Finally, tell students that though ratios and rates can be written in a variety of different forms, the fractional form will probably be the one they use most often because it is an easy form to operate with when applying ratios.

Error Analysis

When writing ratios, students may confuse just which numbers they should be working with. This is especially true when a ratio compares two subsets of a larger set. For example, when asked to compare the number of United Nations members to the number of nonmembers, some students will find the first term of the ratio correctly—185—but will compare it to the total number of countries in the world rather than to the number of nonmembers. Tell students they can avoid such mistakes by first writing out the ratio in word form. For example:

$$\frac{\text{UN member countries} \rightarrow ?}{\text{nonmember countries} \rightarrow ?}$$

Students can then fill in the blanks in the ratio in response to the descriptions. Point out to students that writing the ratio in word form will also help them keep the order of the terms straight.

Follow-Up

REAL-LIFE CONNECTION Ask students what rates we use regularly in everyday life. If necessary, prompt them with some of the following examples. Ask students what terms we generally use when we talk about the following rates: speed (miles per hour, feet per second, and so on), fuel efficiency (miles per gallon, kilowatt-hours per minute, and so on), tire pressure (pounds per square inch), wages (dollars per hour, dollars per year), rental costs (dollars per month, dollars per square foot, and so on).

LESSON 2

Introducing the Lesson

Remind students that on a clock face, every time the second hand makes one complete revolution around the dial, the minute hand moves forward by exactly one tick mark. Write on the board the ratio 1 revolution to 1 tick mark. Ask students how many tick marks the minute hand will move while the second hand makes 45 revolutions.

Write this ratio on the board and urge students to see that these two ratios are equivalent because they express the same relationship—in this case a 1:1 relationship—between second-hand revolutions and minute-hand tick marks. Point out that because students know this relationship, expressed by the ratio on the board, they can identify the number of tick marks the minute hand will move through for any number of second-hand revolutions.

Teaching the Lesson

Because ratios express comparisons, the size of the terms in a ratio is less significant than the relationship between the two terms. One way of making this relationship vivid is to imagine the two quantities being compared on the pans of a balance that has a pointer. If we place 2 items on the left pan and 5 items on the right pan, the pointer will move to a given mark on the scale. If we then double the number of items on each pan, so that there are 4 items on the left pan and 10 on the right, the pointer will still point to the same given mark on the scale. This is because the ratios of the quantities in each case—2 to 5 and 4 to 10—are equivalent.

Tell students that equivalent ratios are most easily worked with when written as fractions. While it may not be immediately obvious to students that the ratios 2 out of every 5 and 4 out of every 10 express the same comparison, the equivalence will become apparent when they place the fractional forms of these ratios next to each other: $\frac{2}{5} = \frac{4}{10}$.

When trying to decide whether two ratios are equivalent, students should first find a common denominator for the ratios. This may involve deciding whether one denominator is already a factor of the other. For example, students should see that when they compare $\frac{2}{5}$ and $\frac{4}{10}$, 10 will be the easiest common denominator with which to work. In other cases, students may have to multiply both denominators to arrive at a common denominator.

Students must be aware that when they multiply or divide to find an equivalent ratio, they must multiply or divide both terms in the ratio by the same number. When the numerator and denominator in $\frac{2}{5}$ are both multiplied by 2, the ratio is in effect multiplied by $\frac{2}{2}$, which is equivalent to 1. Just as multiplying a number by 1 doesn't change the number's value, multiplying a ratio by a fraction equivalent to 1 (in this case $\frac{2}{2}$) doesn't change the ratio's value. The same is true when dividing a ratio by a fraction equivalent to 1.

Error Analysis

When students are asked to write two equivalent ratios for a given ratio, they may come up with many different answers, all of which are equivalent to the original ratio. Be sure they understand that they can use any factor to produce an equivalent ratio but that they must multiply both terms of the original ratio by that same factor. If students seem to be having trouble remembering this, encourage them to write out the factor twice—as both the numerator and the denominator of a fraction—before multiplying. For example:

$$\frac{4 \times 1252}{3 \times 1252} = \frac{?}{?}$$

Follow-Up

CROSS-CURRICULAR CONNECTION Tell students that the notes in musical scales provide a real-life application of equivalent ratios. As they move up an octave in a musical scale, the frequencies of the notes double, but the ratios between the notes stay the same. Tell students that in one octave the ratio of frequency of A to the frequency of C is 220 hertz to 262 hertz (hertz is the number of vibrations per second, or the rate of vibration). Ask them to give the ratio of the frequency of the A to the frequency of the C in the next octave up. (440 hertz to 524 hertz)

Lesson 3

Introducing the Lesson

To help develop students' sense of what simplest form means when working with ratios and their sense of its value, read them the following situation: In its first weekend at the box office, *Mystery* sold 13,000 tickets. An adventure movie, *Drive Fast*, sold 52,000. The ratio of tickets sold for the adventure movie to tickets sold for *Mystery* is 52,000 to 13,000. Write this ratio on the board. Write the ratio 4 to 1 next to it and ask students whether the two ratios are equivalent. As students realize that the two ratios express the same comparison, ask them which form seems to be the more convenient way to express the relationship. Which would be more likely to be used in an advertisement? Discuss the reasons for their choices.

Teaching the Lesson

Generally, it is easier to get a sense of the relationship indicated by a ratio when the ratio is written in simplest form. Tell students that simplest form means, in effect, that the ratio is written using the smallest whole numbers possible. Simplest form also makes it easier to judge whether two ratios are equivalent or to decide which ratio is greater.

Help students see that when they divide a ratio's numerator and denominator by their greatest common factor, they are actually dividing the ratio by a fraction equal to 1. So, for example, when they write $\frac{64}{100}$ as $\frac{16}{25}$, they are dividing $\frac{64}{100}$ by the fraction $\frac{4}{4}$. Remind them that multiplying or dividing a number by 1 does not change the value of the number.

It is not always easy to immediately pick out the greatest common factor of two terms when trying to find simplest form. Remind students that it isn't necessary to use the greatest common factor. Tell them that they can achieve the same end by factoring the terms repeatedly until the terms have no additional common factors. For example, students may quickly see that 64 and 100 have 2 as a common factor. Factoring out this 2 will leave them with the equivalent ratio $\frac{32}{50}$. Help them to see that this ratio is not yet in simplest form because both 32 and 50 have 2 as a common factor. When they factor out this 2 they are left with $\frac{16}{25}$, which *is* the simplest form for the ratio.

Tell students that when they are working with ratios that involve units of measure they must be sure both terms in the ratio use the same measurement unit before they can be sure the ratio is in simplest form. To demonstrate this necessity, have students look at the rectangle on the instruction page. Ask them to explain why the ratio $\frac{1}{6}$ is not helpful when comparing the length of this rectangle to its width. Help them see that such a ratio, written with two different units of measure, can mislead someone into expecting the rectangle's width to be 6 times its length. Point out that such a misconception is not possible when the ratio is written as it should be, as $\frac{12}{6}$ or $\frac{2}{1}$.

Error Analysis

When they are writing ratios in simplest form, students will sometimes choose a common factor that is not the greatest common factor. This error will leave them with a ratio that is equivalent to the given ratio, but that is not in simplest form. For example, a student instructed to write the ratio $\frac{18}{24}$ may factor out a 3 from both terms and write the ratio $\frac{6}{8}$. Tell students that they should always check their answers to see whether the terms of the ratio they have written have any common factors. If they do, students need to simplify further. This is a good time to review basic values of divisibility.

Students who have difficulty simplifying ratios that involve different units of measure may need to review the conversion factors needed to find equivalent units of length, weight, and capacity. If many students encounter such problems, you may want to devote some time to working as a class to develop a chart of conversion factors on the board.

Follow-Up

ADDITIONAL ACTIVITY Explain that the factors used when students rewrite a given measure with a different unit are really rates. For example, there are 4 quarts to every gallon. Ask them what units they could insert to complete the following rates: 5,280 [*feet*] for every [*mile*], 3 [*feet*] for every [*yard*], 16 [*ounces*] for every [*pound*], 16 [*ounces*] for every [*pint*].

LESSON 4

Introducing the Lesson

Tell students that about 50 out of 100 human beings are female and about 50% of human beings are male. Ask students whether the ratio of females is about equal to the ratio of males? Then ask how they could write this ratio in simplest form. Help them to see that both expressions of the ratio $\frac{50}{100}$ are equal to $\frac{1}{2}$.

Teaching the Lesson

Even students who have made the connection between fractions and decimals may initially have difficulty grasping the links among ratios, fractions, decimals, and percents. Try to help them see that a percent is also a ratio—a ratio in which the second term of the comparison is 100 and is implied rather than being explicitly stated. Also point out that just as fractions can be expressed in decimal form, ratios—which can be written as fractions—can also be expressed in decimal form.

Students may find it helpful to have these connections displayed graphically. Write the four terms on the board, spread out at the four compass points. Under each term write the same simple ratio expressed in appropriate form—for example, 1 out of 2, $\frac{1}{2}$, 0.5, and 50%. After you have worked through the instruction portion of the lesson with students, write on the board the algorithms that will change this ratio from one form to the next, using arrows to indicate the direction of the change.

Dividing is only one method for converting fractions to decimal form, and some students may be aware that they can also write an equivalent fraction with a denominator that is a power of ten. You can remind them that they can write a fraction or ratio such as $\frac{3}{8}$ as $\frac{375}{1000}$ (by multiplying numerator and denominator by 125), which can then readily be written in decimal form as 0.375.

You may want to review with students the rules for rounding decimals. Note that when students use a calculator to find a decimal equivalent and a quotient such as 0.6666667 appears on the display, the quotient is already an approximation: in the actual quotient the digit 6 will keep on repeating infinitely. Point out to students that for most of the lessons in this book, it will be sufficient to round decimal equivalents to the nearest hundredth.

Error Analysis

Students may be confused when working with ratios in which the first term is larger than the second term. If so, ask them whether a ratio such as 25:8, when written as a fraction, would be greater than or less than 1. Help them see that if a fraction is greater than 1, its decimal equivalent will also be greater than 1. To help them understand the idea of percents greater than 100, ask them to write the percents for $\frac{98}{100}$, $\frac{99}{100}$, and $\frac{100}{100}$. Once they see that $\frac{100}{100}$ is equivalent to 1, they should understand that 100% is also equivalent to 1, and that a fraction greater than 1 would convert to a percent greater than 100%. Watch for students who try to skirt this topic by reversing the order of the terms and writing, for example, 32% for 25:8.

Follow-Up

CONSUMER CONNECTION Tell students that economic statistics and financial figures are often ratios recorded using one or more of the forms covered in this lesson. Give them the example of the Consumer Price Index (CPI)—the figure the government uses to record inflation, or price increases, at the consumer level. An increase of 3% in the CPI means that consumer prices have increased 3%, on average. Ask students how such an increase in the CPI could be written as a ratio and what units might be attached to it. Help them see that a 3% increase in prices would probably be recorded as a $3 increase for every $100 spent. Give them another example: a sales tax rate of 4 cents on the dollar. Ask them how such a ratio would usually be expressed.

LESSON 5

Introducing the Lesson

A football team has 40 players on its roster and 4 coaches. The ratio of players to coaches is 10 to 1. If the school adds one more coach, does the ratio go up or down? What does the change in the ratio mean for the quality of coaching? Will the change in the ratio give the players more attention from the coaches or less?

Teaching the Lesson

When students begin to compare ratios they must develop a sense of just what it is they are comparing, meaning a sense of what a ratio stands for in a given application and a sense of whether a greater or lesser ratio is more desirable in that situation. You may want to spend some time discussing with students just why it is that a lower ratio of students to computers is better. Writing a ratio in simplest form will help students visualize what the ratio means. For example, when students look at the national average of 12 students to 1 classroom computer, suggest they picture the 12 students clustered around that single computer waiting their turn to use it. Help them see that the situation would be improved (students would have an easier time getting on to the computer) if there were fewer students around the computer. So a ratio of, for example, 6 students to 1 computer would be better. Suggest that they try to get a clear sense of what a ratio stands for before comparing ratios in a problem-solving application.

Comparing ratios can be done by comparing equivalent fractions with a common denominator, but students will find that often in real-life applications, ratios will have denominators that are inconvenient to work with. In these cases, comparing the decimal forms of ratios is generally the easiest method, particularly when a calculator is available.

Error Analysis

As always when working with ratios, students must keep track of the order of terms when they compare ratios taken from applications. Tell them that they may find it helpful first to write the comparison in word form. They can then use the word form as a model when writing the ratios themselves.

Students must also be on the lookout for calculator errors when dividing to find decimal equivalents. They may find it helpful to estimate first, then to compare their calculated answer to their estimate. They can use either of two methods. The first—and simpler—method is simply to estimate whether the decimal will be less than or greater than 1. They can decide this quickly by deciding whether the upper term is less than or greater than the lower term. The second method of estimation would be rounding to compatible numbers. In this way students could quickly see, for example, that the decimal equivalent of $\frac{1500}{135}$ should be close to the value $\frac{1500}{150}$, or 10.

Follow-Up

FAMILY MATH Encourage students who share homes with younger children to find out which child is growing more rapidly by comparing height ratios over time. Discuss the best ratio to use for the comparison. Some students may suggest writing ratios of an older child's height to a younger child's height both now and in the past or now and at some point in the future. Others may suggest that the ratios to be compared should show increase in height over initial height. Ask them what they will first have to do with measurements that are recorded in both feet and inches. Help them see that such measurements must first be written in inches only.

LESSON 6

Introducing the Lesson

Remind students that rates are ratios which compare two quantities that have different measures. Tell them that most often when we use rates in daily life, the value of the second term is 1. For example, when we say a car gets 24 miles per gallon, we use the rate 24 to 1. This makes comparison of rates easy—a car that gets 24 miles per gallon has a better fuel usage rate than a car that gets 23 miles per gallon. Ask the class for other examples of rates in daily life.

Teaching the Lesson

Unit rates are customary in many contexts because they provide a ready means of comparison of rates. Help students see that a common denominator of 1 can make comparison of ratios as direct as comparing two whole numbers. Students should also see, though, that such rates often represent averages rather than assuming an even distribution. For example, the number of traffic deaths in the United States is probably not evenly distributed over all 24 hours of the day. Nor is it evenly distributed over all 365 days of the year. This means that the rate of about 5 traffic deaths per hour represents an averaging out of the total number of traffic deaths each year over the number of hours in a year. Make sure students understand the difference between this sort of rate and a rate that represents a precise and steady quantity, such as the $2 million cost of 1 minute of Super Bowl advertising time.

Point out to students that the method for finding a unit rate is identical to the method for writing a ratio as a decimal: the first term is divided by the second term. Also tell them that unit rates are often written using a shorthand in which the "1" is dropped in the second term and only the unit of measure is used—for example, the unit rate $32.43 per 1 square foot can be written as $\$32.43/\text{ft}^2$.

You may want to explain to students that finding a unit rate by dividing is the same as writing an equivalent ratio with a denominator of 1. In the case of the ratio 14,338.5 gallons/5.5 hours, for example, both terms of the ratio are being divided by the "common factor" of 5.5 to find the number of gallons used in 1 hour.

Error Analysis

With unit rates in particular, students may neglect to write units of measure in their answers. Remind them that all rates compare two measures and that this means that all rates, including unit rates, must be reported with two units of measure.

You may also want to remind students that measures reported with *two* units—2 min 40 sec, for example—must be rewritten using only one unit before the measure can be used to find a unit rate. They should also be aware that when they compare two unit rates, the units of measure used in the terms must be identical. This means, for example, that were they comparing the above rate to a second rate in which one term was 90 sec, they should choose seconds as the unit of measure when rewriting 2 min 40 sec.

Remind students that estimating first will allow them to check the reasonableness of an answer.

Follow-Up

CONSUMER CONNECTION Set up a week-long scavenger hunt in which students go into their local grocery stores and record information from the unit-price tags in an attempt to identify the stores and brands with the lowest prices for given items such as orange juice, rice, and so on.

CROSS-CURRICULAR CONNECTION Ask students to research and compare the per capita incomes for five countries. Have them discuss what per capita income is based on and what it really represents. Ask them what value such a statistic may have when applied to a country in which a very few citizens control the vast majority of the country's wealth.

LESSON 7

Introducing the Lesson

Tell students that in the first months of almost every baseball season, some sportswriters write stories about hitters who are "batting four hundred." Have students discuss what "batting four hundred" means. Ask them if batting four hundred is twice as good as batting two hundred and if it means twice as many hits. Also ask them what it would mean to "bat a thousand" and whether "batting two thousand" would be even better.

Teaching the Lesson

Rather than being reported as percents, success rates and batting averages are customarily reported as decimals to the nearest thousandth. You may want to discuss with students why percents would not be as useful for such statistics. Help them see that the 3 significant digits present in a batting average or success rate give more information than the 2 significant digits reported in a percent. This extra significant digit makes it easier to order such statistics.

You many want to tell students that a player's batting average, or ratio of hits to at-bats, does not account for every situation where a player is at the plate. At-bat figures don't include trips to the plate in which the batter was walked or was struck by a pitch. Also point out that bases reached because of opponents' errors are not counted in hits figures.

Error Analysis

A number of students will likely make the mistake, when finding success rates, of comparing wins to losses rather than wins to total games. Here are three strategies for helping students avoid this mistake. First, suggest a mnemonic that may help them remember which terms to use: When finding success rates, lose the "losses." If students can remember that they shouldn't use the losses figure in their rates, they should be able to puzzle out which figures are used. Second, tell students that a success rate can never be greater than 1 (have students explain why). Also point out that success rates that are close to 1 are quite rare; such an answer should be checked to make sure it incorporates the proper terms. Remind students that both success rates and batting averages are part-to-whole comparisons. Third, if some students are consistently having trouble remembering which terms to use, ask them to always calculate a "failure rate"—such as the rate of losses to games played—when they calculate a success rate. Point out that the success rate plus the failure rate should give a sum of 1 (slightly less if "ties," which are less common, are a possible outcome).

Follow-Up

ADDITIONAL ACTIVITY Give students the following problems concerning batting averages: In the first half of the baseball season, Hank gets 33 hits in 105 times at bat and Babe gets 84 hits in 261 times at bat. Who has the better batting average during the first half of the season? (Babe: 0.322 to 0.314) During the second half of the season, Hank gets 94 hits in 274 times at bat and Babe gets 36 hits in 101 times at bat. Who has the better batting average during the second half of the season? (Babe: 0.356 to 0.343) Predict who has the better batting average for the entire season. Make sure the students do not average the averages. They must combine the stats for both parts of the season and then compute. Now compute the two batting averages. (Hank has the better average for the whole season: 0.335 to 0.331)

DISCUSSION Ask students the following question: At the end of the 1994 season, what percent of the games that he had coached had Pat Riley of the New York Knicks won? (72%) Discuss with students how they could quickly rewrite any success rate to show percent of games won. Make sure they understand that this would generally involve rounding the success rate to the hundredths place.

LESSON 8

Introducing the Lesson

Tell students that the directions on the back of a jar of iced tea mix say to combine two parts water to one part iced tea mix. Ask them how they can use this ratio to mix a batch of iced tea. They should see that they first must determine how much of the mixture they need. Ask what the ratio of water to iced tea mix will be for any amount of water. Help them to see that any ratio they give should be equal to the initial ratio.

Teaching the Lesson

The term "proportion" is often used loosely in everyday parlance to mean "ratio," but it should be used mathematically only when describing a set of two or more ratios that are equivalent to each other. An expression that compares two ratios is a proportion if and only if the two ratios are equivalent. The word proportion itself traces its etymology to the Latin, where it meant "symmetry" or "analogy." Students may find it helpful when remembering what a proportion is to remember the form of an analogy: A is to B as C is to D. Thinking about the original meaning of symmetry may also help students remember that the order of the ratios within a proportion must be the same.

When testing whether two ratios form a proportion, students can use what they already know about testing for equivalent ratios. One method for doing this is finding a denominator that is common to both ratios. Remind students that they may multiply or divide to find this common denominator and that they must perform the same operation on both terms in the ratio (i.e., the numerator and the denominator). Another method for deciding whether two ratios are equivalent is finding their decimal equivalents by dividing the first term in each ratio by the second term. Point out that this second method is particularly easy when students have a calculator handy.

Error Analysis

Because it is essential for students to gain practice with the processes covered in this lesson, their answers to the exercises in this lesson should show more than a "yes" or "no." Suggest that students write out all stages of their work. When they use equivalent ratios, for example, they should write out the original ratio and the number by which they multiplied or divided its terms. With students who seem to be having trouble recording this process, suggest the following form:

$$\frac{4 \times 2}{7 \times 2} = \frac{8}{14}$$

Watch for students who confuse the order of the terms when using a calculator to find a decimal equivalent. Remind these students that the fraction bar in a ratio is really a division symbol, so that a ratio such as $\frac{4}{7}$ can also be written $4 \div 7$.

Follow-Up

ADDITIONAL ACTIVITY Tell students that the ratio of the lengths of the two legs of a right triangle (the legs are the sides adjacent to the right angle) determine the measures of the other two angles in the triangles. This means that if these ratios are equal in two right triangles, the angles in the two triangles are congruent. Have students draw right triangles with legs corresponding to each pair of lengths below. Have them use proportions to identify which right triangles have congruent angles. Then suggest they use a protractor to measure the angles to confirm their findings.

triangle ABC: AB = 3 cm; BC = 4 cm
triangle DEF: DE = 12 cm; EF = 16 cm
triangle GHI: GH = 21 cm; HI = 24 cm

LESSON 9

Introducing the Lesson

Write the ratios $\frac{3}{5}$ and $\frac{6}{?}$ on the board and tell students the two ratios are equivalent. Ask for the missing term in the second ratio. Then ask students to describe how they calculated the missing term. They should see that, because they know the ratios are equivalent, they can find the factor by which the numerator was multiplied and can multiply the denominator by the same factor.

Teaching the Lesson

Students are presented with two methods for finding the value of a missing term in a proportion. Cross-multiplication is the more commonly used method. It has the practical values of being applicable regardless of the compatibility of the terms and of lending itself readily to the use of a calculator. On the other hand, though the number sense method shown in this lesson is useful only when the terms are compatible, when they are compatible this method can be done quickly and without pencil and paper.

When demonstrating the cross-multiplication method for students, you may want to write out the proportion on the board with crossed, double-headed arrows between the ratios connecting the terms to be multiplied. Point out to students that it is this sort of "crossing" that is referred to in the term "cross-multiplication."

The x in the proportion on the first page of this lesson is a variable that represents the missing value. Tell students who are new to the use of variables that any letter can stand in for the missing term here; x and n are common ones. Ask how the x is similar to a ?, ■, or __, which are other placeholders students may have seen before.

You may want to review with students the method for finding the value represented by a variable in an equation. Once students have written the equation based on cross-multiplication, the next step is to isolate the variable on one side of the equation. Then divide both sides of the equation by the factor that shares one side of the equation with the variable. Remind students that they must perform the same operation on both sides of the equation.

Error Analysis

Until students become well-practiced at cross-multiplying, they may make mistakes in setting up the initial equation by multiplying the wrong terms—for example, by multiplying numerator by numerator and denominator by denominator. If such problems appear consistently, suggest that students write out ratios using fraction slashes or colons: 7/20 =x/1800 or 7:20 = x:1800. Then introduce them to the "means/extremes" method of finding cross-products. Point out that the "means" are the middle terms of the proportion, while the "extremes" are the outer terms. Remembering "means/extremes" can help them to remember which terms they need to multiply.

Follow-Up

CROSS-CURRICULAR CONNECTION Introduce students to what is known as the Golden Ratio, which has a value of about 1.61:1. Also known by the Greek letter *phi* (ø), it was revered in many cultures as a mystical or religious number because of its tendency to appear in unexpected places in mathematics. For example, if a line segment is marked off so that the ratio of the length of its longer section to its shorter section is the Golden Ratio, the ratio of the longer section to the entire segment will also be the Golden Ratio. Have groups of students research the Golden Ratio (also known as the Golden Section) and its use in art and architecture through history.

LESSON 10

Introducing the Lesson

Tell students that a certain community center claims that in last year's annual raffle, there were 600 winners out of 7,500 participants. Ask them what information the community center needs to figure out how many winners it can expect to have this year. Help them see that if they know the number of expected participants this year, they can write and solve a proportion.

Teaching the Lesson

As students work the instructional problems in this lesson, they will have to take several steps to ensure that the proportions they write accurately establish the relationships present in the problems. For example, in the second problem, students should see that they can set up two sides of a proportion, one side of which is missing a term. But they must also see that because the term they do have for the second half of the proportion represents the total amount of liquid, they will have to make the bottom term in the first ratio also represent the total amount of liquid. You may want to draw a diagram to help them see this. They should also recognize that they must decide on a consistent unit of measure. Because the cost ratio is given in pints, the pint is probably the easiest unit to use throughout. Finally, students should see that they will need to multiply the number of pints by the cost per pint in order to answer the question posed by the problem. To emphasize the usefulness of proportions as a tool for solving problems, you may want to point out that even this simple multiplication problem can be solved using a proportion ($.89/1 pt = $ x/6 pt).

Error Analysis

Students may find it difficult to set up a proportion correctly from the given information. In most of the exercises here, the information is given in a form that will aid in the writing of a proportion: the two sides are laid out, left and right, and the terms on each side are parallel. Encourage students to pay attention to this form. If they have problems translating it to the form of a proportion, suggest they set up a two-column table for each problem as shown below. The first line in the table will be the ratio they know. They can then rely on multiplication—or even skip-counting—to make subsequent entries.

laundry loads	time
1 load	80 min
2 loads	160 min
3 loads	240 min
4 loads	320 min

Remind them that in the process of identifying the "target" entry—that entry they are working toward to solve the problem—they may have to change units of measure (e.g., hours to minutes).

Follow-Up

ADDITIONAL ACTIVITY Proportions are often used—sometimes correctly, sometimes not—to help make predictions about the future. Tell students that empirical probability is a measure of the chance of something happening in the future based on the circumstances surrounding its having happened in the past. For example, meteorologists look at many different indicators of atmospheric conditions and what sorts of weather have been associated with those indicators in the past. They then make predictions about the chances of specific types of weather happening in the future. Tell students that when a certain set of atmospheric indicators have been present on 430 days in the past, it has rained on 280 of those days. That set of indicators is present today; what is the chance of rain today? (about 65%)

LESSON 11

Introducing the Lesson

Ask students to think about school or yearbook portrait photos. Often such photos are sold in packets that contain prints of the same photo in various sizes. Ask for the ratio of a 4-in. by 5-in. print to an 8-in. by 10-in. print. Help students see that all facial features would be in proportion in the two prints, meaning all comparisons between the two prints would be equivalent to the 1:2 ratio. Then ask whether these features would be proportional to the features of the actual student.

Teaching the Lesson

Very little is constructed in our society without first being planned out by a designer or draftsperson in a scale drawing and\or its three-dimensional counterpart—the scale model. A scale drawing is different from a rough sketch in that every dimension in the drawing is related to the life-size dimensions according to a single ratio, known as the scale. This precision and proportionality allows the designer or drafter to see precisely how parts will fit together in the actual project. It also allows those who work on the actual project to place a ruler on the drawing itself in order to find a specific life-size dimension.

Tell students that the scale in any scale drawing is simply a ratio—usually in simplest form—of the drawing measure to the actual measure. As such, the scale can be used to form a proportion to calculate any actual measure from its corresponding drawing measure.

Point out that often in a scale the two terms have different units of measure. Students should be aware that they don't need to change to a single unit of measure as long as they use the same units in both ratios when setting up a proportion. For example, if a scale ratio shows inches over yards, the ratio on the other side of the proportion must also show inches over yards. The missing term can then be calculated directly in the appropriate unit of measure.

As students work through exercises on their own, they may invert the scale ratio when finding a missing measure. You may want to point out that the ratio can be ordered either way as long as both ratios in the proportion have the same orientation.

Error Analysis

It is essential that the orientation of scale measure to actual measure be the same for both ratios when students set up their proportions. Students who invert one of the ratios can catch this error themselves if they first decide whether the missing measure should be greater than or less than its counterpart in the other ratio. For example, when students look at the first ratio in the lesson, $\frac{1}{4}$ in./5 yd = 3 in./x yd, they should be able to judge that because the upper term is larger in the second ratio, the lower term should also be larger in the second ratio. This means they should expect an answer that is considerably greater than 5 yd. If they end up with an answer that is less than 5 yd, they can assume they have probably inverted one of the ratios.

Also watch for students who confuse units of measure when cross-multiplying. Suggest that after they have read and understood the problem, they write out the form of the answer using the proper units and leaving a blank that they can fill in once they have calculated the missing value.

Follow-Up

CROSS-CURRICULAR CONNECTION Tell students that lenses that make things appear larger or smaller create scale images. Have them identify the scale in an image seen through a microscope that enlarges images 400 times. Ask them to calculate the actual size of a paramecium that appears to be 2.5 mm in length when viewed through this microscope.

LESSON 12

Introducing the Lesson

Have students sketch out maps of the route to the nearest building exit. Ask them whether their maps are roughly to scale—that is, whether the distances on their maps are related to each other in a way similar to the way in which the actual distances are related to each other. Discuss with them why this is a useful convention in mapmaking.

Teaching the Lesson

Ocean navigators have historically plotted courses by taking their measures directly from maps using a pair of hand-held calipers. It is vitally important that such maps be consistent in their application of the map scale. Tell students that this also holds true for road maps, hiking maps, and so forth; misrepresentations of size get people lost. In this way, every map is really a scale drawing of the mapped region.

Many maps have the scale given not only numerically, but also graphically, so that the user can measure out the distance in question and can then physically move the measuring standard to the scale to compare it to the mile or kilometer markings there.

Because the map scales in this lesson make use of fractions and mixed numbers, students will have to divide by fractions to find the missing measures in some of these exercises. Dividing by a fraction gives the same result as multiplying by its reciprocal. The algorithms in this lesson condense the division step in solving the equations and show only the multiplication by the reciprocal. You may want to break out the process even further and connect it to the solution process they have used before. Show that, for example, when they solve the equation $\frac{3}{2}\times n = 5$ they are still dividing by $\frac{3}{2}$ to isolate n.

Error Analysis

In addition to the customary difficulties of working with proportions, students working through these exercises may also have trouble when dividing by fractions. One common error is multiplying by the fraction itself rather than by the reciprocal of the fraction. Students can guard against such errors by using number sense and deciding before they multiply whether their answer should be greater than or less than the original number. For example, if $\frac{3}{2}$ of n is 5, n itself must be less than 5. An answer that is greater than 5 probably indicates that the student failed to find the reciprocal before multiplying.

Follow-Up

CAREER CONNECTION Cartographers have always been faced with the problem of how best to project a spherical world onto a two-dimensional surface. Numerous projections have been proposed and many of these are in use in specific areas. Have students research the various projections and the benefits and drawbacks of each. They should find that some are preferable for preserving distance, some for preserving relative land area, and so on. Ask students which type of projection might be most appropriate or desirable for each of the following uses: a sailor's navigation maps, a map showing the decline of forest land worldwide, a map showing the islands that surround the North Pole.

LESSON 13

Introducing the Lesson

Using a straightedge, draw a large triangle on the board and label its vertices A, B, and C. Now draw a line segment YZ in it that is parallel to BC and that has endpoints on AB and AC. Ask students to identify as many relationships as they can between triangle ABC and triangle AYZ. They should immediately see that both triangles share at least one congruent angle—angle A. Help them to see that the angle AYZ is congruent to angle ABC and that angle AZY is congruent to angle ACB. Have a volunteer come to the board with a yardstick to measure and record the lengths of the sides of the triangles. Ask students whether they can see any specific relationships among the ratios of side lengths between the two triangles.

Teaching the Lesson

To help students develop a better sense of just what is meant by "same shape" when defining similarity, you can suggest they trace triangle DEF on a sheet of paper and cut it out. They can then place it on top of triangle ABC and try to center it within the larger triangle. This should help give them a sense that the smaller triangle is a direct reduction of the larger. To help quantify what this means, they can slide the cut-out triangle to match it up, angle for angle, with the larger one. This will strengthen their sense both of what is meant by "corresponding angles" and of the fact that corresponding angles are congruent in similar triangles. As they match up angles, they should also make the connection with the triangle you drew on the board in the introductory activity above.

Before students work through the first exercise, you may want to point out that the orientation of similar triangles is not always the same: for example, triangles ABC and XYZ in Exercise 1 are rotated in relation to each other. Students should see, though, that according to the test for similarity that they will use in this lesson—congruent corresponding angles and proportional corresponding sides—these rotated triangles are indeed similar. Application of this test should also convince them that flipped triangles can be similar.

Error Analysis

When they test triangles for similarity, students may have trouble correctly identifying corresponding sides. Such difficulty can lead them to identify two similar triangles as dissimilar. Suggest that before students check for proportionality they write down the letters of the first triangle. Students can then, either by visual estimation or the use of other measuring tools, pair each angle with a corresponding angle from the second triangle. If the context for comparison warrants it, students can use tracing paper to orient one triangle with another triangle.

Follow-Up

DISCUSSION Pose the following problem: Can you draw two triangles that have all angles congruent that are *not* similar?

You may want to have volunteers come to the board to try. Help students to see that, once they have a second triangle with congruent angles, trying to change the length of one side must change either the lengths of both the other sides proportionally, or the measures to two of the angles. Students should conclude that identifying three pairs of congruent angles in triangles guarantees they are similar. Once they do this, challenge them to decide whether identifying two pairs of congruent angles is enough to guarantee similarity.

LESSON 14

Introducing the Lesson

Ask whether any student knows the approximate distance from the earth to the sun (the mean distance is about 93 million miles). Discuss with students just how scientists might find a measurement that they obviously cannot measure directly (in this case, by using parallax and trigonometry). Help students see that such distances can be measured indirectly by using other known measures.

Teaching the Lesson

Indirect measurement involves measuring objects or distances—usually large objects or distances—by making use of other known measurements and applying the special relationships present in similar triangles or trigonometric ratios. The initial goal in identifying which known measurements can allow you to indirectly take an unknown measurement is to be able to draw similar triangles, often similar triangles that share a vertex created by two intersecting lines. Any pair of intersecting lines forms two pairs of congruent angles. For example, in the lesson diagram that illustrates indirect measurement across a river, lines AE and BD intersect at point C. This means that angles ACB and ECD are congruent, as are angles ACD and ECB. Each pair of angles is also known as a pair of vertical angles, and students should intuitively recognize that they are congruent.

Because any two pairs of congruent angles are enough to guarantee that two triangles are similar, once you have formed a pair of congruent vertical angles, you need only form one other pair of congruent angles. This other pair is often a pair of right angles. Once these angles have established two similar triangles, you can make use of the proportionality of corresponding sides to find missing measures.

Other forms of indirect measurement take advantage of the fact that the ratio of object height to object shadow is identical for all local objects at any given time of day.

Error Analysis

One of the difficulties students may encounter when finding measurements from the diagrams in this lesson is difficulty in identifying corresponding sides in the similar triangles. Tell students that when they are working from a diagram that includes two triangles that share a vertex (such as the river diagram), triangle sides that lie along the same line are corresponding sides. This means that the sides that are farthest apart (sides that are parallel) are also corresponding sides.

Follow-Up

ADDITIONAL ACTIVITY Challenge students to draw their own maps to make an indirect measurement. Read aloud the following problem, giving students time to complete each step before you read the next.

Joan has been swimming laps across Crater Lake all summer. Every day she swims from the dock to the lookout tower on the other side. How can you help her estimate this distance using what you know about similar triangles?

(a) The line from the dock to the tower runs due west. At the dock, Joan uses a compass to find due north. She walks north for 100 meters, then puts a marker stick in the ground. How would you draw this on your map?
(b) She keeps walking north another 10 meters and marks the spot with a stone. Then she turns due east and starts walking again. What should your map look like now?
(c) As she walks east, she keeps looking toward the marker stick and the tower. When the two are exactly aligned, she stops. She has walked east 22 meters from the stone. How will you add this to your map?
(d) Where are the two similar triangles on your map?
(e) What is the approximate distance across the lake from the dock to the lookout tower? (220 m)

LESSON 15

Introducing the Lesson

On the board draw two similar triangles, ABC and EFG. Tell students the triangles are similar and ask them to identify the corresponding sides and angles. Then make the triangles part of similar quadrilaterals by attaching another triangle to each. Label the new vertices D and H. Ask students whether these new triangles are similar to each other. Then ask students what they can conclude about the corresponding angles and corresponding sides in the quadrilaterals ABCD and EFGH. They should see that the angles are congruent, and they may identify corresponding sides as being proportional.

Teaching the Lesson

The definition of "similar" that was introduced earlier with triangles can be generalized to all plane figures: similar figures are figures that have the same shape but not necessarily the same size. By this definition, scale drawings are similar to the life-sized items they represent. All circles are also similar to one another. When looking at simple polygons, though, students can test for similarity by applying the same tests they apply to triangles. If corresponding angles are congruent and corresponding side lengths are proportional, the polygons are congruent. Thus students should see that the quadrilaterals you drew on the board in the introductory activity are similar quadrilaterals.

With triangles, students concluded that two pairs of congruent angles were enough to guarantee similarity. When they are dealing with special types of quadrilaterals, they can also narrow down the number of tests that they must perform to check for similarity. By the end of this lesson they should conclude that all squares are similar. They should also see that when checking rectangles they need only test for proportionality of sides, though they must use adjacent sides. What students may or may not conclude from this lesson is that if the corresponding angles in other types of quadrilaterals are congruent, the quadrilaterals must be congruent unless they are trapezoids. This is because the parallel sides in trapezoids can be shortened or lengthened without changing the measures of the angles.

Error Analysis

As when working with similar triangles, students may encounter difficulties identifying corresponding sides when setting up a proportion to test for similarity. This can be even more difficult when dealing with quadrilaterals, especially quadrilaterals that are rotated or flipped with respect to each other. Again, suggest students first write out the vertex letters of each figure with the letters of the corresponding angles of the second figure beneath them. When quadrilateral names are written one beneath the other this way, students should see that any line segment they take from the first set of vertex letters must correspond to the line segment immediately above or below it.

Follow-Up

ADDITIONAL ACTIVITY Challenge students to extend their work with similarity to other types of polygons. Remind them that regular polygons are polygons with congruent angles and congruent sides. Have them draw several regular hexagons and octagons and decide whether all regular hexagons and octagons are similar. Then ask whether they can generalize this observation to apply to all regular polygons. Have them write out a formal statement of this rule.

LESSON 16

Introducing the Lesson

Give students the following problem: An aluminum fabricator is rolling small sheets of aluminum into cylinders to make tennis ball cans.

Each can must hold three balls, and each of the balls is $2\frac{3}{4}$ in. in diameter. The sheets of aluminum are $8\frac{1}{2}$ in. by $9\frac{1}{2}$ in. rectangles. Which way should these sheets be rolled to form the cans—along the long dimension or along the short dimension?

Encourage students to think about how long the can must be and then to cut, roll, and tape a sheet of notebook paper. Ask whether any students were surprised that the shorter dimension was the height of the can. Discuss about how many times the diameter of the ball must be multiplied to find the length of a sheet that will wrap all the way around it.

Teaching the Lesson

The circumference of a circle is loosely defined as the distance around the circle and is roughly analogous to the perimeter of a polygon. This connection may be easier to envision if you imagine a regular polygon inscribed inside a circle so that all of its vertices are on the circle. The greater the number of sides the inscribed polygon has, the closer its perimeter will be to the circumference of the circle.

Students who have trouble envisioning what circumference is and how it can be measured may be helped by thinking about tightening a tape measure around a can or by thinking about cutting the can down its side, flattening it out, and measuring its length.

Emphasize to students that the ratio of circumference to diameter is exactly the same for all circles, regardless of the size of the circle. There is no decimal or fraction that expresses this ratio directly, so we use the Greek letter *pi* (π) to stand for this value. However, we need to use some value for this ratio when computing size; an approximate value of the ratio is 3.14 to 1, which means that a circle's circumference is about 3.14 times the length of its diameter. Point out that the value $\frac{22}{7}$, which is also just an approximation of π, may be easier to compute with when the diameter has a fraction in it.

Error Analysis

Suggest that students first estimate circumference so that they will be able to test the reasonableness of a calculator answer.

If students seem to be having difficulty understanding the concept of the circumference/diameter ratio, you may want to have them measure a variety of circular objects for themselves. They can use a tape measure for this, or they can use a string, which they can then hold up to a yardstick. Suggest that they record their results in a table like the one below and that they round their quotients to the nearest hundredth. When they have made sufficient entries and are surer of the relationship, ask them to describe how they could use this information to find the circumference of a circular item with a diameter of 10 cm.

Object	Circumference (C)	Diameter (d)	$C \div d$
tire	18 in.	$5\frac{3}{4}$ in.	3.13

Follow-Up

ADDITIONAL ACTIVITY Ask students how they could rewrite the formula $C = \pi d$ to use it to find the diameter when they know the circumference. They should recognize that to isolate d on one side of the equation, they would have to divide both sides by d, which would give a formula $d = \frac{C}{\pi}$. Once they have this formula, ask them to apply it. Have them measure the circumferences of circular classroom objects and predict their diameters.

Lesson 17

Introducing the Lesson

On the board, draw a 4-by-4 grid of 16 equal squares. Within that grid inscribe a circle that has a diameter of 4 squares. Now ask students to estimate the area covered by the circle. They should see that the area of the 4-by-4 grid is 16 square units, and that the area of the circle is about 12 units. You may want to divide the squares into halves and quarters so that students can refine their estimate as much as possible. Have them identify the radius of the circle (2 units) and relate their estimate of the circle's area to its radius (the area is about 6 times the radius).

Teaching the Lesson

Whereas a circle's circumference is directly proportional to its radius or diameter, a circle's area is proportional to the square of its radius. By the end of the lesson, students should see that this means that doubling a circle's radius or diameter quadruples its area. This relationship is stated in the formula $A = \pi r^2$, which is a formula students will encounter regularly in future math classes.

In this lesson, because the focus is on comparison of radii ratios to area ratios, students will report many of their answers using π as one of the terms. You may want to point out that such reporting is perfectly valid—π represents a precise value—unless the problem at hand calls for a numerical value as an answer. An area reported using π is in fact more precise than an area reported using an approximated value of π. Reporting answers with π also better illustrates the proportional relationship between a circle's radius and its area.

Error Analysis

Students probably have not had much experience reporting answers with non-numeric terms, so some may fail to do so when working through these exercises, either by substituting some value for π when it is inappropriate, or dropping π from the answer altogether. Remind these students that because π is in the original formula, it must be present in some form in the computed answer. Point out that when an approximate value such as 3.14 is substituted for π, this value is then a factor in the computed answer. The term π itself can also appear as a factor in a computed answer.

Watch for students who mistakenly substitute the circumference formula when they should be finding the area, or who substitute the value of the diameter into the formula instead of the value of the radius. Encourage these students to copy out the formula $A = \pi r^2$ every time they start a new problem.

Also watch for students who neglect to report their answers using square units. Point back to the figure you drew on the board for the introductory activity above and remind them that area describes the number of square units enclosed by a boundary.

Follow-Up

CONSUMER CONNECTION Ask students to describe how they might go about solving the following problem: Ray's Pizza and Calzone King both sell pizza slices for $1.00 apiece. Ray's slices come from a 12-slice pie that is 12 inches in diameter. Calzone King's slices come from an 8-slice pie that is 10 inches in diameter. Which pizza parlor has the better unit rate (the most pizza per dollar)? Tell students that they can find the answer without substituting an approximate value for π. (Ray's: 3π in.^{2}; Calzone King: 3.125π in.^{2})

LESSON 18

Introducing the Lesson

Describe the following situation to students: Galway wants to buy carpeting for his living room, which is 13 ft by 17 ft. He multiplies these dimensions and finds he needs 221 square feet of carpeting. When he gets to the carpet showroom, he realizes carpeting is sold by the square yard, not by the square foot. He knows there are 3 feet to the yard, so he divides 221 by 3 to get about 74 square yards of carpeting. He chooses a style he likes and places his order for 74 square yards, amazed at how much it costs him to carpet his living room. When the carpet dealer delivers the product to his home, Galway discovers he has enough to carpet the entire house and part of the back yard. How did he misjudge so badly the amount he needed?

Suggest that students think about this individually for a moment, perhaps drawing pictures if that helps. Eventually some students will come to realize that even though there are 3 feet to a yard, there are 9 square feet to a square yard; Galway should have divided the number of square feet by 9 instead of by 3.

Teaching the Lesson

The general formula for the area of a rectangle is Area = length × width, or $A = l \times w$. If students need to be reminded of how this formula is derived, mark out a rectangle on the floor tiles (or draw a grid on the board and mark out an area). Show students that the length of the rectangle equals the number of tiles in one row and the width equals the number of rows. The total number of tiles (the area) can be found by multiplication.

The formula for the area of a square is a special case of the rectangle area formula. Given that by definition the length and width of a square are congruent, we can substitute as follows: Area of a square = length × width = side × side = side^2, or s^2.

Just as the area of a circle increases as the square of its radius increases, so the area of a square increases as the square of a side increases. Students who understood the concept as applied to circles should readily grasp it here when it is applied to squares.

Error Analysis

Watch for students who, when finding the ratio of the areas of two squares based on the ratio of their side lengths, forget to square the side-length ratio. Suggest that these students draw grids within the squares in question to get a better sense of the factor by which the area increases.

Follow-Up

CONSUMER CONNECTION Give students the following problem to complete in small groups: Jed has funds for only 100 m of fencing. He wants to use this fencing as efficiently as possible by creating the largest rectangular playground he can with it. He creates a chart to show combinations of rectangle width and length using only 100 m of fencing. Complete the chart.

length (m):	5	10	15	20	25	30	35	40	45
width (m):	45	(40)	(35)	(30)	(25)	(20)	(15)	(10)	(5)

Now have students use grid paper to create a graph that shows length along the bottom axis and area along the side axis. Have them compute the area for each of the rectangles in their chart and graph it. When they have graphed all 9 points, ask them what type of rectangle makes for the most efficient use of fencing. They should see that a square gives the greatest area for a given perimeter. Ask them to speculate about whether a circle would be even more efficient than a square. They may wish to perform some computations to check on their predictions.

LESSON 19

Introducing the Lesson

Fold and tape a piece of posterboard into a 1-inch cube. Bring in a larger box—at least as large as a shoebox—and pass both it and the 1-inch cube around the classroom. Ask students to estimate the number of such 1-inch cubes it would take to completely fill the box. Then ask how, using the single 1-inch cube as a model, they could develop a plan for systematically deciding the number of such cubes that would be required to fill up the box. Lead them to see that if they covered the bottom of the box with 1-inch cubes, the number of such cubes would be equal to the area of the bottom face. Then encourage them to see that that number (the area) would have to be multiplied by the number of layers of cubes in the box. Record this formula on the board as area $\times$ height and ask what the formula for area of a rectangle is.

Teaching the Lesson

The general formula for the volume of a rectangular prism (a solid in which all faces are rectangles) is Volume = area of base $\times$ height. A cube is a special type of rectangular prism that allows us to condense this formula. The base in a cube is a square, so the area of the base is equal to side $\times$ side, or side^2. The height is also equal to the side, so the special formula for the volume of a cube becomes Volume = side^2 $\times$ side, or side^3, or s^3.

Students who are having difficulty understanding the concept of powers and exponents will need a review here. Write 5^3 on the board and label the base (5) and the exponent (3). Remind students that just as 6×4 is shorthand for 6 used as an addend 4 times $(6+6+6+6)$, 5^3 is shorthand for 5 used as a factor 3 times $(5 \times 5 \times 5)$. It is also important that students develop a sense that using a number as a factor three times means the product will increase rapidly and will likely be much greater than the original number.

Point out to students that writing and cubing ratios is the more efficient way to answer the questions in Exercise 3. They should see this is so after they take the time to compute the volumes.

Error Analysis

Students who seem to be having trouble understanding the relationship between the side length and the volume of a cube should work with unit cubes or dice to build larger cubes. Ask them how many smaller cubes it would take to build a larger cube that was three unit cubes along each edge. Suggest they build the cube a layer at a time, counting and recording the cubes in each layer as they go along.

Follow-Up

ADDITIONAL ACTIVITY During the twentieth century, an inventor named Buckminster Fuller pursued the idea that the basic shape for the average house, a cube, was less efficient than a house built in the shape of a hemisphere, or dome. One of his arguments was that a dome had the best ratio of enclosed space to surface area. If you wish, show that this is mathematically correct (formulas will be too difficult for most students to work with). Have students do outside research on Fuller. Then, in class, discuss his ideas about efficient housing. Are they practical?

ANSWER KEY

LESSON 1 (pages 2–3)

1 a. 20:3
b. $\frac{20}{3}$
2 a. $\frac{2}{3}$
b. $\frac{3}{5}$
3. 12:5, $\frac{12}{5}$
4. 2:7, $\frac{2}{7}$
5. 5:18, $\frac{5}{18}$
6. 5,000:750, $\frac{5{,}000}{750}$
7. 2:1, $\frac{2}{1}$
8. 3:2, $\frac{3}{2}$
9. a. $\frac{7}{3}$
b. $\frac{3}{10}$
c. $\frac{7}{13}$
d. $\frac{10}{20}$
10. a. $\frac{13}{45}$
b. $\frac{20}{78}$
c. $\frac{78}{45}$
11. Answers will vary.
12. Answers will vary.

LESSON 2 (pages 4–5)

1. a. $\frac{21}{9}$
b. $\frac{21}{9}$
2. a. $\frac{6}{2}$
b. $\frac{12}{4.5}$, $\frac{6 \times 2}{2 \times 2} = \frac{12}{4}$
c. No
3–6. Answers may vary.
3. $\frac{2}{4}$, $\frac{3}{6}$
4. $\frac{8}{6}$, $\frac{12}{9}$
5. $\frac{6}{16}$, $\frac{9}{24}$
6. $\frac{1}{20}$, $\frac{2}{40}$
7. yes
8. no
9. yes
10. no
11. 30 pairs of sneakers, 20 pairs of oxfords, 10 pairs of loafers
12. Answers will vary.
13. Answers will vary.

LESSON 3 (pages 6–7)

1. b. $\frac{12}{5}$
2. b. 75¢
c. $\frac{50\text{¢}}{75\text{¢}} = \frac{2}{3}$
3. $\frac{1}{4}$
4. $\frac{176}{1}$
5. $\frac{6}{5}$
6. $\frac{3}{1{,}000}$
7. $\frac{2}{1}$
8. $\frac{3}{32}$
9. $\frac{3}{20}$
10. $\frac{2}{1}$
11. $\frac{3}{1}$
12. $\frac{4}{1}$
13. $\frac{1}{2}$
14. yes
15. The ratio remains the same because each new fold doubles both the number of shaded regions and the number of unshaded regions.

LESSON 4 (pages 8–9)

1. a. 1.44
b. 1.44 = 144%
2. 3.13, 313%
3. 0.57, 57%
4. 0.24, 24%
5. 0.4, 40%
6. 0.9, 90%
7. 0.33, 33%
8. 0.04, 4%
9. 5.71, 571%
10. 0.34, 34%
11. $\frac{49}{100}$
12. $\frac{1}{5}$
13. $\frac{37}{100}$
14. $\frac{7}{25}$
15. $\frac{3}{20}$
16. $\frac{3}{4}$
17. $\frac{9}{20}$
18. $\frac{1}{4}$
19. $\frac{91}{500}$
20. Yes, they all express the same ratio; $\frac{12}{25} = .48 = 48\%$
21. Atlanta 87%, Savannah 5%, Columbus 3%, Cleveland 3%, 4 soccer sites 2%
22. 25%

LESSON 5 (pages 10–11)

1. a. 3
b. $\frac{3 \times 2}{4 \times 2} = \frac{6}{8}$
c. $\frac{7}{8} > \frac{6}{8}$
2. a. .38, .09
b. .38 > .09
3. $\frac{3}{9} > \frac{20}{63}$
4. $\frac{2}{7} < \frac{1}{2}$
5. $\frac{1}{8} < \frac{1}{4}$
6. $\frac{1}{10} > \frac{1}{100}$
7. 0.14 > 0.13
8. 0.63 < 0.69
9. 1.5 < 1.8
10. 0.92 > 0.91
11. 3 L to 3 mL
12. 4 min to 180 min
13. 5 yr to 90 days
14. 5 to 6, 11 to 12, 4 to 3

15. **$\frac{1}{5}$, $\frac{1}{13}$, $\frac{1}{31}$** When the numerator equals one, the larger the denominator, the smaller the fraction.

LESSON 6 (pages 12–15)
- **1. b.** 25
- **2. b.** $1.19, $1.49
  - **c.** $1.49 $>$ $1.19
- **3.** 55.2 words/min
- **4.** 17.8 pages/sec
- **5.** 74.6 customers/hr
- **6.** 0.98 TV sets/household
- **7.** 23.1 blinks/min
- **8.** 3.75 c/min, or 0.06 c/sec
- **9.** 0.2 gal/sec
- **10.** 17.6 yd/sec
- **11.** $2.87/page
- **12.** $1.20/gal
- **13.** $1,722.25/yr
- **14.** $933.13/student
- **15.** $4,595/hour
- **16.** $.43/oz
- **17.** $>$
- **18.** $>$
- **19.** $<$
- **20.** $\frac{5}{11,298} > \frac{3}{8,748}$
- **21.** about 2 hours
- **22.** $2,499.90
- **23.** Answers will vary.

LESSON 7 (pages 16–17)
- **1. a.** 175, $\frac{98}{175}$
  - **b.** 98 $\div$ 175 = 0.56
  - **c.** 0.56
- **2. b.** 27, 0.293
- **3.** 0.588
- **4.** 0.136
- **5.** 0.475
- **6.** 0.325
- **7.** 0.545
- **8.** 0
- **9.** 0.316
- **10.** 0.222
- **11.** 0.255
- **12.** 0.273
- **13.** 0.341595, 6; 0.344208, 3; 0.342065, 5; 0.344407, 1; 0.344338, 2; 0.342125, 4

LESSON 8 (pages 18–19)
- **2. b.** $>$, $\neq$
  - **c.** do not
- **3.** yes
- **4.** yes
- **5.** no
- **6.** yes
- **7.** yes
- **8.** no
- **9.** yes
- **10.** no
- **11.** yes
- **12.** yes
- **13.** no
- **14.** no
- **15.** yes
- **16.** no
- **17.** yes
- **18.** no
- **19.** Yes, 1 $\times$ 3 = 3, and 18 $\times$ 3 = 54

LESSON 9 (pages 20–23)
- **1. a.** 3
  - **b.** 4
- **2. a.** 5
  - **b.** 2, y=17.5
- **3.** 15
- **4.** 9
- **5.** 4
- **6.** 32
- **7.** 6
- **8.** 12
- **9.** 44
- **10.** 7
- **11.** 30
- **12.** 20
- **13.** 1.5
- **14.** 5
- **15.** 1
- **16.** 3
- **17.** 108
- **18.** 3.5
- **19.** 5
- **20.** 1
- **21.** 5 in.
- **22.** $7.92
- **23.** 300 mi
- **24.** $41\frac{2}{3}$ hr
- **25.** 1225
- **26.** 42
- **27.** 5 c
- **28.** 3.5 min
- **29.** 367.5 in.
- **30. a.** 9 qt
  - **b.** 2 qt
  - **c.** yellow, 3 gal; red, 1 gal
  - **d.** $4.99

LESSON 10 (pages 24–27)
- **1. a.** 9
  - **b.** $3.75
- **2. a.** 3
  - **b.** 51, y=25.5 grams
- **3.** 6 loads
- **4.** $9.30

5. 115.5 mi
6. $3\frac{1}{2}$ hr
7. 1050 times
8. 6 CDs
9. 250 mg
10. 465 mg
11. 96 persons
12. 2 pt
13. 1000 cars
14. $2,187.50
15. 3,500
16. 440 points
17. 29 tickets
18. 120 in.
19. 15 fluid ounces
20. 60 g
21. 1.04 in.
22. 7.2 cm
23. 320 mi
24. .031 in.
25. 2.88 days

LESSON 11 (pages 28–31)

1. **a.** $\frac{n \text{ in.}}{20 \text{ ft}}$
   b. $\frac{1}{2} \times 20 = 4 \times n$; $10 = 4 \times n$; $\frac{10}{4} = n$; 2.5 in. $= n$
2. 300 ft
3. 225 ft
4. 15 ft × 290 ft
5. 100 ft × 250 ft
6. 25 ft apart
7. 0.3 in.
8. 0.4 in.
9. 1.1 in.
10. 3 in.
11. 2 ft
12. 0.9375 in.
13. 0.792 in.
14. Answers will vary.

LESSON 12 (pages 32–35)

1. **c.** $\frac{4}{3}$, $\frac{20}{12}$, $\frac{5}{3}$ mi, $1\frac{2}{3}$ mi
2. 6.25 mi
3. 47.5 mi
4. 33.75 mi
5. 13.125 mi
6. 20 mi
7. $53\frac{1}{3}$ mi
8. 40 mi
9. 60 mi
10. 30 mi
11. $53\frac{1}{3}$ mi
12. 60 mi
13. Iditarod
14. Unalakleet
15. Answers will vary.

LESSON 13 (pages 36–37)

1. **a.** $\frac{5}{15}$, or $\frac{1}{3}$
2. $6 \times a = 4 \times 9$, $6 \times a = 36$, divide by 6, $a = 6$
3. yes
4. no
5. $a = 6.5$, $b = 5$
6. $x = 16$, $y = 4.5$
7. Answers may vary. Ratios of corresponding sides and angles are equal.

LESSON 14 (pages 38–41)

1. **a.** 168
   b. 105
2. **b.** $16n = 48{,}000$, divide by 16, $n = 3{,}000$
3. 18
4. 72
5. 9 ft
6. 60 ft
7. 60 ft
8. 200 ft
9. Answers will vary.

LESSON 15 (pages 42–43)

1. **a.** yes
   d. yes
   e. yes
2. yes
3. yes
4. Yes. Angles are all 90° and the ratio of side lengths is always 1:1.

LESSON 16 (pages 44–45)

1. **c.** 25.8 in.
2. 201 yd
3. 36.7 mm
4. 1,203.25 m
5. 31.85 mm
6. 13.4 ft
7. 5.9 m
8. 100.48 cm
9. 3.85 mi

LESSON 17 (pages 46–47)

1. **b.** 36π cm^2
   c. $\frac{1}{9}$
   d. 3, 9, 3
2. 16π cm^2
3. 49π in.^{2}
4. 25π mm^2
5. 121π m^2
6. First way: Double the given radius, square it, and multiply by π. Second way: Multiply the area found using the given radius by 2^2 or 4.
7. 50.24 cm^2, 153.86 in.^{2}, 78.5 mm^2, 379.94 m^2
8. The intial area (100π mi^2) times 25 gives an

area of 2,500π mi^2. The new radius is the square root of 2,500, so the new range is 50 miles.

LESSON 18 (pages 48–49)

1. b. $\frac{900 \text{ ft}}{100 \text{ ft}} = \frac{9}{1}$
c. 3
2. $\frac{4}{1}$, $\frac{16}{1}$, 16
3. $\frac{1}{8}$, $\frac{1}{64}$, 6400
4. one

LESSON 19 (pages 50–51)

1. a. 64,000 m^3
2. a. $\frac{1}{4}$
b. $\frac{1}{64}$
c. 64
3. 1000
4. 512,000 in.^{3}
5. 343 mm^3
6. 512 in.^{3}
7. 74,088 mm^3
8. a. 600 in.^{2}, 1,000 in.^{3}
b. 9, 27

CUMULATIVE REVIEW: LESSONS 1–4 (page 52)

1. $\frac{7}{1}$, 7:1
2. $\frac{25}{5}$, 25:5
3. $\frac{8}{3}$
4. $\frac{3}{8}$
5. $\frac{2}{5}$
6. $\frac{6}{1}$
7–9. Answers will vary.
7. $\frac{2}{8}$, $\frac{3}{12}$
8. $\frac{10}{14}$, $\frac{15}{21}$
9. $\frac{1}{5}$, $\frac{4}{20}$
10. $\frac{3}{5}$, yes
11. $\frac{12}{7}$, $\frac{18}{7}$, no
12. $\frac{15}{40}$, yes
13. no
14. 0.67, 67%
15. 0.3, 30%
16. 1.8, 180%
17. $\frac{22}{25}$
18. $\frac{1}{20}$
19. $\frac{9}{25}$
20. $\frac{1}{8}$

CUMULATIVE REVIEW: LESSONS 5–7 (pages 53)

1. $<$
2. $>$
3. $>$
4. $>$
5. $>$
6. $<$
7. $>$
8. $>$
9. $<$
10. 12 cm to 30 m
11. 8 hr to 60 min
12. 4 exams/hr
13. $890,411,000/day
14. $<$
15. $<$
16. 0.587
17. 0.387
18. 50 million

CUMULATIVE REVIEW: LESSONS 8–10 (pages 54)

1. yes
2. yes
3. no
4. no
5. no
6. yes
7. no
8. no
9. no
10. yes
11. yes
12. no
13. 6
14. 36
15. 3
16. 55
17. 8
18. 70
19. 12.5
20. 1
21. 21 walruses
22. 0.25 qt
23. $.80
24. 9 cookies
25. cross-multiplication and number sense

CUMULATIVE REVIEW: LESSONS 11–12 (page 55)

1. 2 ft
2. 3 ft
3. 6 in., or $\frac{1}{2}$ ft
4. 2 × 1 ft
5. yes
6. 13 mi
7. 6 mi
8. 32 mi
9. a, b: map would have to be very large to include the whole state; c: Metric is generally not used for distances in the United States; d is correct

CUMULATIVE REVIEW: LESSONS 13–15 (pages 56)

1. $A = 8$, $B = 22.5$
2. 40 m
3. yes

4. no
5. The length of your shadow, your own height, and the ray of sunlight which causes your shadow are lengths of a triangle that is similar to a triangle formed by the same ray of sunlight, the canyon's shadow, and the one mile distance across.

CUMULATIVE REVIEW: LESSONS 16–19 (pages 57)

1. 94.2 cm
2. 3.5 in.
3. 28.9 m
4. 200.96 square in.
5. 706.5 square in.
6. 25,434 square mm
7. $\frac{3}{1}$, $\frac{9}{1}$, 9
8. $\frac{1}{6}$, $\frac{1}{36}$, 900
9. The area quadruples
10. 125
11. 512
12. 15,625 cubic m
13. 8 cubic ft
14. 125 cubic m
15. 4,096 cubic ft